无处不在的龙

Dragons Everywhere

Gunter Pauli

[比] 冈特·鲍利　著
[哥伦] 凯瑟琳娜·巴赫　绘
田　烁　王菁菁　译

上海远東出版社

特别感谢以下热心人士对童书工作的支持：

匡志强　宋小华　解　东　厉　云　李　婧　庞英元
李　阳　刘　丹　冯家宝　熊彩虹　罗淑怡　旷　婉
杨　荣　刘学振　何圣霖　廖清州　谭燕宁　王　征
李　杰　韦小宏　欧　亮　陈强林　陈　果　寿颖慧
罗　佳　傅　俊　白永喆　戴　虹

目录

Contents

老虎很想知道龙是否真的存在。他的朋友鲸坚定地相信龙仍然生活在我们身边，就藏在地球某个神秘的地方。

“我知道你们鲸是世界上最大的动物，但是如果在附近的深山中还有龙存在，那你们就不是世界上最大的动物了。”老虎指出。

A tiger is wondering out loud if dragons really exist. His friend, the whale, firmly believes that there are still dragons around, hiding in secret places on the planet.

"I know you whales are the biggest animals in the world, but if there are still some dragons living in the mountains around here, then you are not the biggest animal after all," the tiger points out.

老虎很想知道龙是否真的存在

A tiger is wondering out loud
if dragons really exist

蜻蜓

The dragonfly

“嗯，我们很久很久以前生活在陆地上，也用四肢走路，但是现在生活在海洋中，也就看不到谁是最大的动物了。”鲸回答。

“没错，”老虎说道，“我想知道龙有没有翅膀。”

“我知道的唯一有翅膀的龙是蜻蜓。”

“Well, ages ago we used to live on land and walk on all fours as well, but now that we’re living in the sea, there is no way to see who is the biggest anyway,” replies the whale.

“True,” says the tiger and then says, “I wonder if dragons have wings.”

“The only dragon I know of that has wings is the dragonfly.”

“但那不是龙啊！他是一种会飞的昆虫。”老虎反驳道。

“好吧，他确实是昆虫。不过是一种特殊的昆虫，因为他有四个翅膀，大一些的在后面，小一些的在前面。但是，请告诉我，他为什么被叫作蜻蜓——会飞的龙呢？”鲸问道。

“But that’s no dragon! It is a fly,” protests the tiger.

“Well, an insect, at least. And an exceptional one, with its four wings. The bigger ones are at the back and smaller ones in front. But tell me, why do you think it’s called a dragonfly then?” the whale asks.

他是一种会飞的昆虫

It is a fly

对于这么小的生物来说，这简直是神速啊

Amazing speed for a creature so small

“因为他们长着一对大眼睛，很吓人。”

“但是，他们只吃其他昆虫，比如蚊子、黄蜂和蜜蜂。他们从来不伤害人类。你知道吗？他们每小时能飞将近100千米呢。”

“真的吗？对于这么小的生物来说，这简直是神速啊！但他们也还算不上是龙啊，即使那些20厘米长的大个头蜻蜓也不算。还有，他们也不会喷火呀。”

“People may find them scary because of their big eyes.”

“But they only eat other insects, such as mosquitos, wasps, and bees. They never hurt people. Did you know that they can fly at nearly a hundred kilometres an hour?”

“Really? Amazing speed for a creature so small! But they still don’t qualify as dragons, not even the big ones that grow up to twenty centimetres long. And they don’t breathe fire either.”

“深海龙鱼也算不上，尽管他们能发出很亮的光。”鲸说。

“是的，我听说过。当你潜入海洋深处时，一定能看到他们发出的亮光。可是，这些光也算不上是火啊！”

“Nor do the dragonfish found here in the ocean, although they do have some very bright lights,” says the whale.

“Oh yes, I’ve heard of those. You must have spotted their flashing lights when diving in the depths of the ocean. Their lights still aren’t fire, though!”

深海龙鱼

Dragonfish found here in the ocean

我们应该小心的是那些科莫多龙

What we should be worried about is the Komodo dragons

“我确实碰到过这些聪明的小鱼，”鲸回答，“你知道吗？当他们吃饱时，就不再发光了。雄性深海龙鱼长着大大的牙齿，但个头甚至比蜻蜓还要小，所以没什么好怕的。”

“我们应该小心的是那些科莫多龙。”老虎提醒道。

“是的，他们不光长得像龙，而且皮肤上长满了鳞片，摸起来也很像龙呢。还有啊，他们可臭啦！就像真的龙一样臭。”

“I have indeed come across these clever little fish,” replies the whale. “Did you know that when their stomachs are full, they turn off their lights? The males have very big teeth, but as a dragonfish is even smaller than a dragonfly, there is nothing to worry about.”

“What we should be worried about is the Komodo dragon,” warns the tiger.

“Yes, they don’t only look like dragons; their scaly skin probably feels like dragon skin too. And boy, do they stink! Just like real dragons.”

“你是怎么知道这些的？”老虎笑道，“我听说，科莫多龙甚至可以杀掉水牛呢。”

“当科莫多龙发起攻击时，谁都不能保证自己的安全。他们可以长到3米长，体重达到70千克！”

“没错，一定要小心，尽管他们的听力不那么好，但是他们的视力特别好，可以看到几百米以外的猎物。”

“How would you know that, anyway?” laughs the tiger. “I’ve heard Komodo dragons can even kill water buffalo.”

“No one is safe when a crush of Komodo dragons attack. They can grow as long as three metres and weigh seventy kilos!”

“Right. But do be careful. While their hearing is not that good, they do have sharp eyesight and can spot prey hundreds of metres away.”

他们的视力特别好

They do have sharp eyesight

雷龙之地

The land of the Thunder Dragon

“好吧，很高兴知道这些，谢谢你的提醒。但是我住在海洋里，是安全的，倒是你应该小心了。我想我们可以得出结论了，世界上只有科莫多龙存在，再也不可能存在真的龙了。”

“嗯，这样的话，我就要去不丹旅行了，那可是雷龙之国。在国王的统治之下，那里的人们似乎是世界上最幸福的。”

“如果你想和那里的巨龙作伴，那你真应该在那里生活，但我会想念你的，我的朋友。”鲸感叹道。

“Well, that's good to know. Thanks for the advice, but as I live in the ocean, I am safe. It is you who should take care. I guess we can conclude that there are only Komodo dragons and no true dragons to be found anymore.”

“Well, in that case I will travel to Bhutan, the land of the Thunder Dragon. Under their king's reign, the people there seem to be the happiest lot in the world.”

“If that is the grand dragon you want to be with, then his land is the place you should live. I will miss you though, my friend,” sighs the whale.

“我也会想你的，”老虎回答，“但我还是要去安全的地方。我已经厌倦了现在被人类围猎的生活了，他们想要我的皮和骨头。”

“我相信不丹人还有他们的国王会欢迎你的，因为你雄壮而优雅，你将会在那里安全、快乐地生活。再见吧，我亲爱的朋友。”

“再见！”

……这仅仅是开始！……

“I will miss you too,” replies the tiger, “but I have to go where I’ll be safe. I am tired of being hunted by people who want my skin and bones.”

“I am sure that the people of Bhutan, and their king, will celebrate you for your power and grace, and that you will be safe and happy there. Farewell then, dear friend.”

“Farewell!”

... AND IT HAS ONLY JUST BEGUN!...

……这仅仅是开始！……

... AND IT HAS ONLY JUST BEGUN! ...

Did You Know?

你知道吗？

Dragonflies only flap their wings 30 times per minute, compared to 600 times for a mosquito and 1,000 times for a fly. The dragonfly has 20 times more power in its wings than any other insect.

蜻蜓每分钟仅扇动翅膀30次，而蚊子要扇动600次，苍蝇1000次。蜻蜓翅膀的力量要比其他任何昆虫大20倍以上。

The dragonfly lives most of its life as a nymph. It flies for only a fraction of its life. A dragonfly is able to see 360° around and 80% of its brainpower is dedicated to vision.

蜻蜓一生大部分时间是若虫，只有很少的一段时间是可以飞行的。蜻蜓的视野范围能达到360度，其80%的脑力都被用来发展视力。

Komodo dragons are the largest lizards on Earth. They are dominant predators and prey on deer, pigs and even water buffalos. Any prey animal wounded by them will soon die of blood poisoning as a result of the many strains of bacteria present in the Komodo dragon's mouth.

科莫多龙是地球上最大的蜥蜴，它们是占支配地位的食肉动物，以鹿、猪甚至水牛为猎物。科莫多龙的嘴里含有大量细菌，任何被它们咬住的猎物都会因为血液中毒而很快死掉。

A Komodo dragon can eat 80% of its weight in a single feeding. Local people call the Komodo dragon the "land crocodile", as it has large, serrated teeth with which it tears up its prey.

科莫多龙一次可以吃掉的食物重量可以达到自身体重的 80%。当地人将科莫多龙称作“陆地鳄鱼”，因为它有巨大的锯齿形牙齿用来撕碎猎物。

A female Komodo dragons can reproduce without having had a male fertilise her eggs.

雌性科莫多龙不需要雄性为它的卵授精，可以自己繁殖。

Dragons are mythical creatures featured in the myths, legends, and folklore of many cultures. When cartographers placed dragons on a map, it was to indicate dangerous or unexplored territories.

在很多文化里，龙是神话和民间传说中的神秘生物。当绘图者在地图上标记出龙的图标时，就表明这个地方是危险的或未知的土地。

Ancient people who found dinosaur fossils mistakenly assumed they were the remains of dragons.

发现恐龙化石的古代人错误地将其认定为龙的遗迹。

One of the symbols of Bhutan is the Thunder Dragon or Druk (in Dzongkha). Bhutan is also known as Druk Yul or The Land of Druk. Bhutanese leaders are known as Druk Gyalpo or Thunder Dragon Kings.

不丹国的一大象征是雷龙，也因此被称为“雷龙之国”，他们的的领袖被称作“雷龙国王”。

Do you think that real dragons ever existed?

你认为存在过真正的龙吗?

Why is it that people use the word "dragon" in different animal names such as dragonfly, dragonfish, Komodo dragon and blue dragon, knowing that dragons do not exist?

为什么人们明知龙不存在，却用"龙"这个字来给其他动物命名呢?比如dragonfly（蜻蜓）、dragonfish（深海龙鱼）、Komodo dragon（科莫多龙）和blue dragon（蓝龙）。

As tigers are endangered, do you think they need to live in a country where they are safe?

老虎已经濒临灭绝，你认为它们应该去一个安全的国家生活吗?

Would you like to live in a country where the king is considered a dragon?

你愿意生活在一个国王被看作是龙的国家吗?

Do It Yourself!

自己动手!

Make a list of all the animal names you can find that have "dragon" as part of them. Then make a list of all the famous stories and films (old and new) about dragons. Does it seem to you as if there are dragons everywhere and that they have been around since the beginning of time?

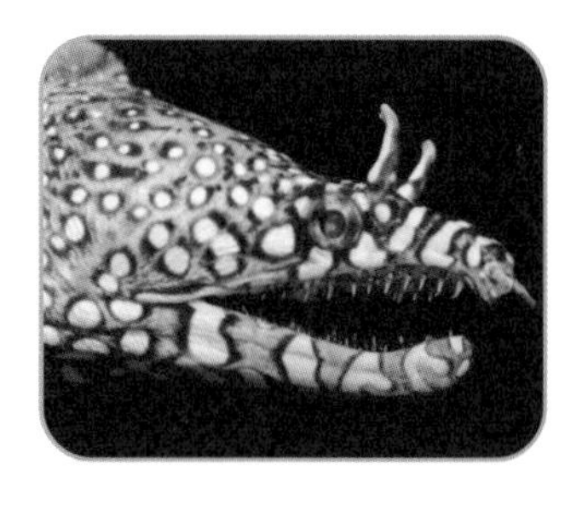

列出你知道的所有含有“龙”字的动物名字，再写出所有关于龙的著名故事和电影（不管新旧）。你是不是觉得龙好像是无处不在的，始终就在我们周围?

学科知识

Academic Knowledge

生物学	鲸在进化中是独一无二的，因为它离开海洋去陆地生活过，后来又回到了海洋；蓝鲸是地球上有史以来最大的动物；蜻蜓没有牙齿，但有很强的下颌骨来嚼碎猎物；若虫是一些无脊椎动物尚未成熟的形态；若虫与幼虫的区别；深海龙鱼是一种生物性发光的深海鱼；雌性深海龙鱼（40厘米长）与雄性（5厘米长）身材大小的区别。
化　学	生物性发光的主要化学反应剂是荧光素（具有发光性蛋白质），它能在辅酶因子（如钙和镁）的作用下产生氧化作用。
物　理	生物性发光中的光是由生物的发光器产生的。
工程学	蜻蜓可以达到很快的飞行速度，可以像直升机一样悬停，像蜂鸟一样倒飞，也可以直上直下飞行，每分钟仅需扇动翅膀30次。
经济学	特有物种（比如鲸、科莫多龙和老虎）对游客有很强的吸引力，因而拥有很大的经济价值；不丹用国民幸福总值（GNH）来衡量经济增长幅度与国民幸福程度。
伦理学	老虎已濒临灭绝，怎么能允许捕杀它们，又怎么能允许老虎被当作马戏团的动物来供人娱乐呢？我们是在和其他物种共享地球还是只是地球上所有物种的主宰者？
历　史	蜻蜓在地球上已经生活了3.25亿年；在公元前4世纪，一位历史学家在中国四川省发现剑龙的化石并将其误认为龙；在宗教改革和理性时代到来后，龙已经变成了一个娱乐性话题，而不再含有精神意义；自公元9世纪开始，龙的标志在威尔士一直占有主导地位，1859年还出现在了威尔士国旗上。
地　理	印度尼西亚的科莫多岛；深海龙鱼生活在大西洋和墨西哥湾1 000—3 000米深的地方，被称作是海洋中层区至深海区；不丹国旗采用了中国风格的龙图腾，龙的四爪还攥有珍珠；马耳他共和国的国旗是圣乔治杀龙的图腾。
数　学	红蜻蜓数学挑战赛是一项解决系列数学问题的比赛。
生活方式	神话在我们生活中的重要性；我们如何既接受古老的神话，又欣赏新的神话中龙扮演的重要角色，比如《霍比特人》和《哈利波特》。
社会学	龙和蜻蜓的神话；武士将蜻蜓视为力量、敏捷和胜利的象征；中国人将蜻蜓与繁荣、和谐联系在一起；美洲原住民将蜻蜓视为幸福、速度和纯洁的象征。
心理学	在自我实现的过程中，蜻蜓象征着变化——情感更加成熟，对生活的理解更加深入，散发着力量和高贵优雅的风度。
系统论	一些物种（如老虎）的自然保护区已经很少了，不丹是为数不多的能为老虎提供理想生存环境的国家，由于文化、习俗、神秘主义和实用主义的综合因素，不丹为老虎的存续提供了安全的环境。

情感智慧

Emotional Intelligence

老　虎

老虎在幻想着那些假设情况下的问题，探索着创造性思维的极限。然而，在和鲸的对话中，他变得非常实用主义，开始探究含有“龙”字的动物名字。他系统地观察了几个不同地区的动物，得出了结论：他要到雷龙之国去生活，那里的人们非常幸福。他终于找到避难之处，可以摆脱与偷猎者的持续斗争。最后，他设想自己的后代能有一个安全的未来。

鲸

鲸喜欢老虎发挥想象力的方式，并分享了自己的经历：他曾经是生活在陆地上的哺乳动物，用四肢来行走，后来又再一次变回了海洋哺乳动物，就无法比较自己和龙的体型大小了。鲸沉浸于列举带有“龙”字的动物名字，分享着他知道的一切。他们最后得出结论——龙并不存在。当老虎分享了自己厌倦了被追猎的日子，并打算迁往幸福之地生活的愿望时，鲸鼓励他去一个安全的地方，并安慰老虎说在那里他将会被当作珍稀物种而受人们爱护。

艺术

The Arts

“龙”是一种最具代表性的幻想艺术形象。几千年来，艺术家、作家为这种惊人、古老、强大的生物所着迷。龙的传说和人类文明一样悠久。它们有些被描绘为有翅膀的，还有一些被描绘为能喷火吐冰的庞然大物。现在该由你来描绘自己心目中龙的形象了。你既可以让它是友好的，又可以让它是残忍的，也可以让它有翅膀或没有翅膀。最后将你和朋友们描绘的龙的形象放到一起展示一下。

思维拓展

Systems: Making the Connections

龙的传说和人类的文明史一样悠久，是伴随着人们采掘珍贵金属和建筑石料的足迹演化而来的。基于人们对动物解剖学的掌握程度，奇特、巨大、未知的动物化石遗骨被一片一片地拼凑在一起，通过这些拼凑物，人们想象这些物种曾经真实存在过，这很可能是那些关于奇异的蜥蜴类生物——龙的传说产生的原因。无论古代还是现代，不同的文化都信奉这些传说，还有很多人认为这是真的。引人注目却又濒临灭绝的现代物种，比如鲸和老虎，数量已经急剧下降，有一天可能也会像传说中的龙一样消失，除非我们意识到应该保护濒危物种，不再猎杀它们，侵占它们的栖息地。过去的几十年，人们试图通过建立国家公园和立法的方式来保护濒危物种，通过在它们的栖息地巡逻来驱赶偷猎者。实践证明这些方法是失败的，因为我们并没有让犀牛和大象免遭偷猎者的杀害。我们要为这些濒危物种创造理想的生存环境，让它们继续进化之路，以更大程度地自由享受生命进程。世界上只有少数几个国家有能力、有空间、有文化氛围和财力资源，为像老虎这样的大型猫科动物提供可以自由漫步的自然环境。不丹最初并不被认为是老虎的家园，但后来隐藏的摄像机拍摄到了上百只老虎自由漫步的情景。然而，不丹并不想发展所谓的“老虎观光游”，而是试着找到平衡点，既能让老虎自由生活，又能满足农民需求，保障他们依赖居住地资源谋生的权利，同时妥善处理老虎对公共安全所带来的威胁。他们的方式与众不同，并且启发人们用一种更全面系统的方式来解决问题，而不是在经济收入和环境保护中只取其一。

动手能力

Capacity to Implement

对于保护老虎，你能作出怎样的贡献？你想提出哪些建议?老虎和人类有没有和谐共存的可能？老虎有没有繁衍下去的可能？有没有一些方法，让人们既能利用老虎来挣钱，又不用活在被老虎攻击的恐惧之中？和你的家人朋友讨论一下这些问题。反思一下，要想实施你的方法，法律法规应该作出哪些修改?

故事灵感来自

This Fable Is Inspired by

不丹王太后

HM the Queen Mother of Bhutan

不丹王太后是不丹第四世国王吉格梅·辛格·旺楚克的第一位妻子。她曾在印度西孟加拉邦噶伦堡的圣约瑟夫修道院和印度古尔塞翁圣海伦学院接受教育。她为改善所有民众，尤其是不丹偏远地区民众的生活质量而发起了众多项目，对环境保护和建立国家公园、生物长廊有着浓厚的兴趣，在国际活动中经常强调削减贫困、保护环境和女性社会角色转型的重要性。她还是一位文学热爱者，也是一位很有成就的作家，著有《彩虹与云》《雷龙的宝藏——不丹国的肖像》等作品。

图书在版编目(CIP)数据

冈特生态童书.第三辑修订版：全36册：汉英对照 /
(比)冈特·鲍利著;(哥伦)凯瑟琳娜·巴赫绘；
何家振等译.—上海：上海远东出版社，2022
书名原文：Gunter's Fables
ISBN 978-7-5476-1850-9

Ⅰ.①冈… Ⅱ.①冈… ②凯… ③何… Ⅲ.①生态环境-环境保护-儿童读物—汉、英 Ⅳ.①X171.1-49

中国版本图书馆CIP数据核字(2022)第163904号

著作权合同登记号图字09-2022-0637号

策　　划 张　蓉
责任编辑 程云琦
封面设计 魏　来李　廉

冈特生态童书
无处不在的龙
[比]冈特·鲍利　著
[哥伦]凯瑟琳娜·巴赫　绘
田　烁　王菁菁　译